The Nature Kid's Guide to
PANDAS

Level 2

RENATA MARIE

LP Media Inc. Publishing
Text copyright © 2023 by LP Media Inc.
All rights reserved.

For information address LP Media Inc. Publishing,
3178 253rd Ave. NW, Isanti, MN 55040
www.lpmedia.org

Publication Data

Pandas
The Nature Kid's Guide to Pandas — First edition.

Summary: "Learn all about Pandas, the Nature Kid Way"
— Provided by publisher.

ISBN: 978-1-954288-58-4

[1. Panda Bears – Non-Fiction] I. Title.

Title: The Nature Kid's Guide to Pandas

CONTENTS

UP IN THE MOUNTAINS

Mist falls over a forest in China. Up in a tree, a big ball of fur rolls over. It is a giant panda.

The panda's habitat is wet. It rains and snows. Sometimes a cloud hangs low over the trees. **The panda's fur keeps it dry.**

Below, the forest is thick with bamboo. The panda climbs down the tree with its sharp claws. It has a long day of eating and sleeping ahead.

Pandas live up to 10,000 feet (3,048 meters) high in the mountains.

ROLY-POLY PANDAS

The panda moves slowly through the forest. He has thick legs. His body is round.

Pandas have tiny tails. They tuck them close to their bodies. Their small ears can hear faraway sounds. **Their noses can smell faraway pandas.**

Not many animals hunt adult pandas. They are big. They can be three feet (0.9 m) tall at their shoulders.

GIANT EATERS

The panda bites into bamboo. He has large teeth. His strong jaws squash the plant.

Pandas have long wrist bones. It acts like a thumb. They use it to hold their food.

Pandas are mainly **herbivores**. They mostly eat bamboo. Bamboo is a plant, but panda stomachs are made to eat meat. **Eating plants is hard on their stomachs.**

Bamboo does not give pandas a lot of energy. They need to eat a lot of bamboo to get the strength they need. Pandas spend 12 hours each day eating.

SNIFFING FOR SNACKS

The panda sniffs for food. Bamboo is easier for him to find than meat.

Sometimes pandas cannot find bamboo. They have to eat other food. They eat eggs. They hunt small animals.

Sometimes they go to a farm. They eat beans, pumpkins, and wheat. **They even like pig food.**

They look for food, water, and safe places to sleep. They drink melting snow.

Bamboo
Eggs
Beans
Pumpkins
Small animals
Wheat
Pig food
11

12

Snow falls on the black and white bear.

In winter, most bears sleep in dens. They are safe. **Not pandas.** Eating bamboo does not give the Panda enough fat to sleep all winter.

Pandas need to stay safe. So they **camouflage**. They need dark fur in summer and light fur in winter. But shedding uses a lot of **energy**.

Shedding uses a lot of energy. Pandas cannot change their fur twice a year. Instead, they shed all the time.

BLACK AND WHITE

The panda's thick white fur makes him hard to see in the snow. It also keeps him warm in the cold.

Pandas can hide in summer too. Pandas have dark markings on their bodies. They have marks around their eyes. **They have markings across their backs.** They have black ears and legs. The black fur helps them hide in the shade of trees.

Giant pandas can weigh over 300 pounds (136 kilograms)

Panda fur feels wooly like sheep fur.

SOUND SLEEPERS

**Cozy in a tree,
the panda falls asleep.**

It is good that pandas can hide. They almost never run. **And they take a lot of naps.**

Pandas sleep in nests and trees. They sleep in caves. They sleep on the forest floor. They will sleep almost anywhere.

When pandas find a spot, they curl up.
Or they lie on their side, belly, or back.

DAY OR NIGHT

The moon shines over the forest. The panda wakes up.

Some animals sleep at night. Others sleep during the day. **But pandas do both.** Pandas are up during the day. And they are up at night.

Pandas can see in the dark. Their eyes look like cat eyes. This helps them look for food at night.

DO THE DOO

**The panda eats.
He sleeps.
And he poops.**

Pandas cannot break down bamboo easily. A lot of what they eat will come out of them later.

Breaking down so much food is tiring. Between each meal, pandas sleep. They nap for two to four hours. They even poop in their sleep.

Pandas can poop over 40 times a day. That is a lot of bamboo!

PEE UP
A TREE

The panda also pees—upside down. Slowly, he uses his back legs to climb up a tree. And then he pees.

The pee leaves a smell on the tree. **Male pandas hope female pandas will smell the pee.** They want to make more pandas.

Male pandas have other ways of marking their path. They claw or rub against trees. They call out into the forest for a female to hear.

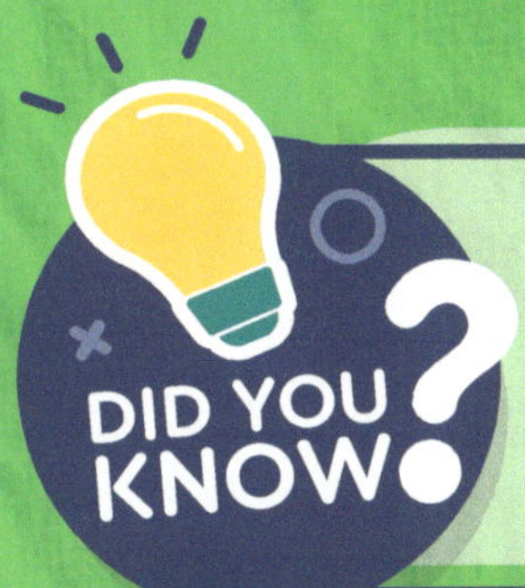

Pandas can travel up to 6.2 miles (10 kilometer) in one day.

CUTE CUBS

A female panda sleeps in a ball. She holds a baby panda in her arms.

Baby pandas are called cubs. Pandas have one to two cubs at a time.

Cubs are small. **They also cannot see or hear.** They need their mothers to keep them safe. It would take 900 cubs to be the same size as a female adult panda.

FUN FACT!
Newborn cubs only weigh three to five ounces (85 to 142 grams).

BABY
BEARS

For the first month, the mother panda keeps her cub close.

After one week, cubs grow black and white markings. After six to eight weeks, they open their eyes. **At five months, they finally learn to walk and climb.**

Pandas are **mammals**. They drink milk when they are young. Mother pandas will nurse their cubs until they are nine months old. At six months, a cub will start eating bamboo.

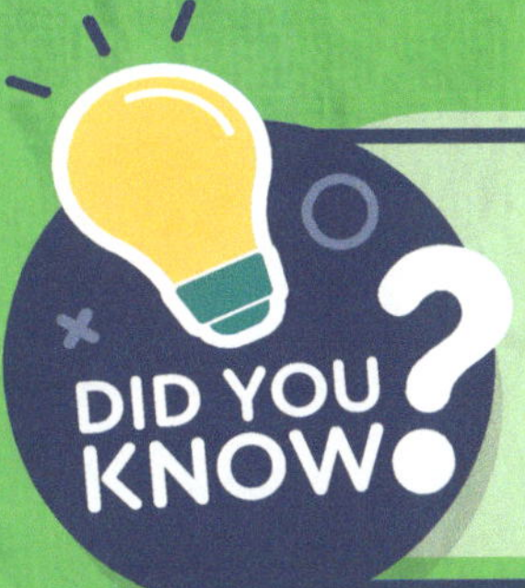

Cubs learn to climb by climbing on their mothers.

PLAYFUL
PANDAS

Two balls of fur play under the trees. The cubs roll and wrestle.

They swim in streams. They chase each other up trees.

Pandas are quiet and live alone. **Sometimes they run into other pandas.** Then, they like to talk.

When cubs are two years old, they will find their own home in the mountains.

As cubs get older, they will learn new sounds.

MAMA BEAR

The mother panda is too big for most animals to hunt. But her cub is small.

Snow leopards, jackals, and yellow-throated martens hunt cubs.

Do not mess with a mother bear. Pandas are strong, and their jaws do more than break bamboo. Pandas have one of the strongest bites of land mammals.

But hungry animals are not the only dangers to pandas.

Snow leopard
Jackal
Yellow-throated
marten

A SMALLER WORLD

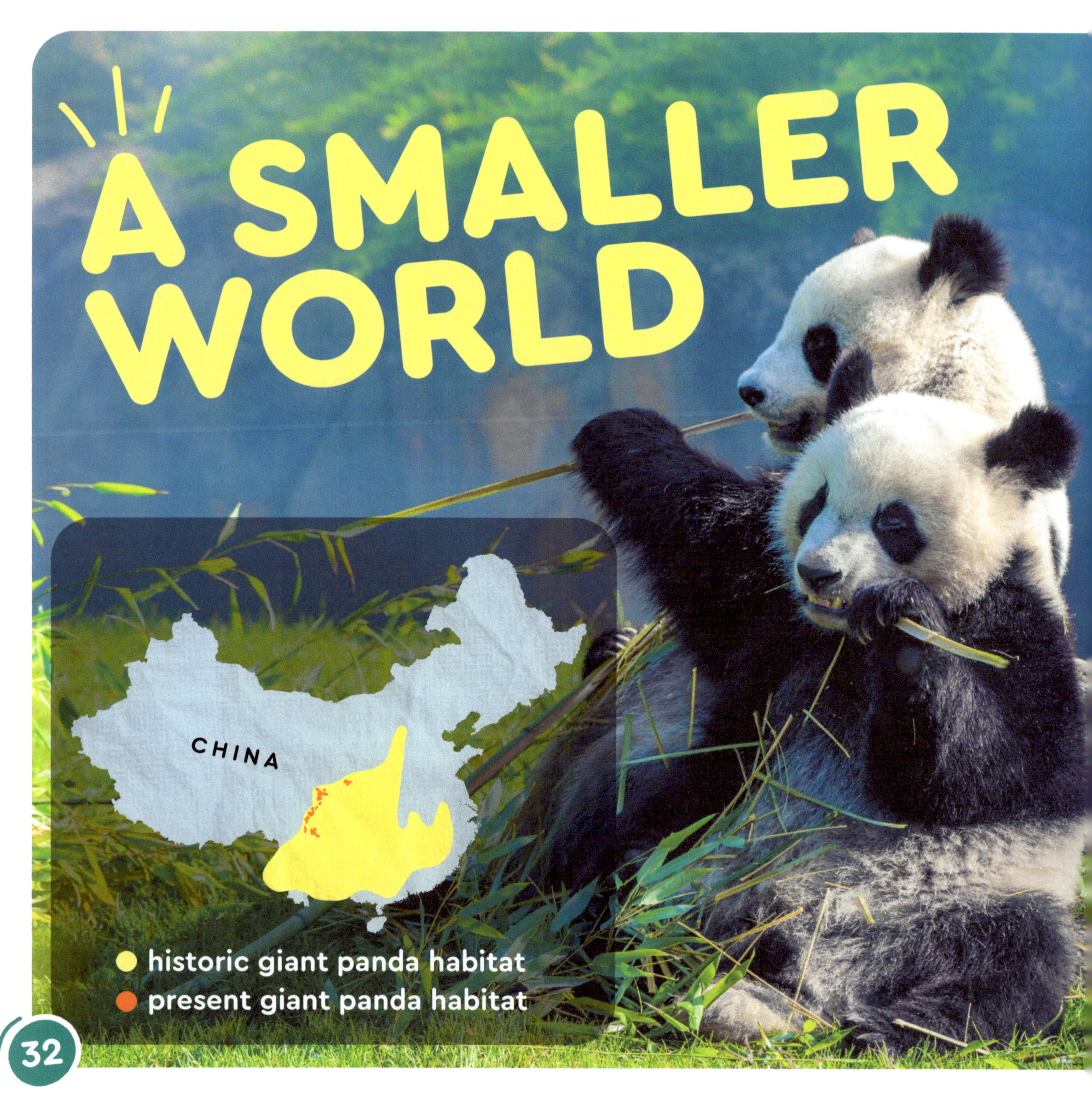

Pandas used to live in large parts of Asia. Now there are only 1,850 giant pandas in the wild.

Pandas are a **vulnerable** species.

Why are there so few pandas?

Bamboo takes years to grow. Normally, a panda would travel to find new bamboo. But humans moved in.

They broke up the forest. They cut down trees. They built farms. They built roads. And pandas could no longer find new plants. Less bamboo means fewer pandas.

Pandas had fewer cubs. Males could not reach the females. The remaining pandas had to live in the mountains.

MOVING
HOMES,

A panda climbs up a mountain. He is shy. He does not want to run into the humans who live below.

The mountains are becoming hotter. Earth is warming. **It is too hot for bamboo to grow.** The plant can only grow higher in the mountains where it is cooler.

Pandas have to move higher up the mountain to find food. The tops of the mountains are smaller. The forests are smaller. There is less bamboo. And the pandas are hungry.

KIDS CAN HELP THE EARTH

HELPFUL HUMANS

People are trying to help pandas.

The Chinese government is helping to grow forests. It is bridging forests so pandas can travel. It made parks for pandas to safely live in. **And it made hunting pandas illegal.**

Three hundred pandas live in zoos. Zookeepers help pandas make more pandas. They take care of cubs. And they give them a safe place to live.

China
made it
ILLEGAL
to hunt
pandas

YUMMY
TREATS

A panda sits in a zoo. Kids watch her play with a bamboo puzzle. She finds the food inside!

In zoos, pandas play with toys and eat healthy food. They eat bamboo. They also eat carrots, apples, and sweet potatoes.

At the zoo, people keep pandas safe. They hope that one day, pandas will once again live peacefully in the forests.

Pandas live for 15 to 20 years in the wild. They live about 30 years in zoos.

GLOSSARY

herbivores
animals that eat plants
page 9

camouflage
something that helps an
animal hide in its habitat
page 13

mammals
animals that feed
their babies milk.
Humans are mammals.
page 27

energy
the strength to do things
page 13

vulnerable
in danger
page 33

habitat
where an animal lives
page 5

MORE AMAZING ANIMAL BOOKS
from Nature Kids Publishing!

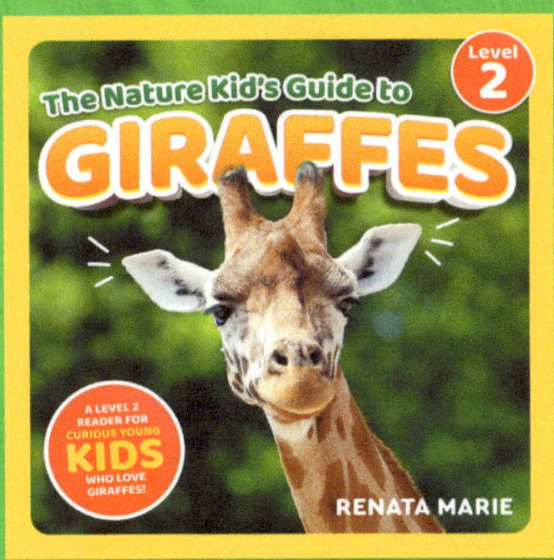

Giraffes

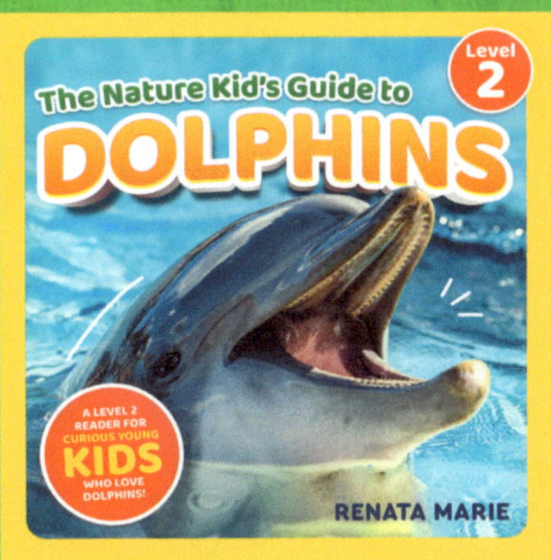

Dolphins

Rabbits

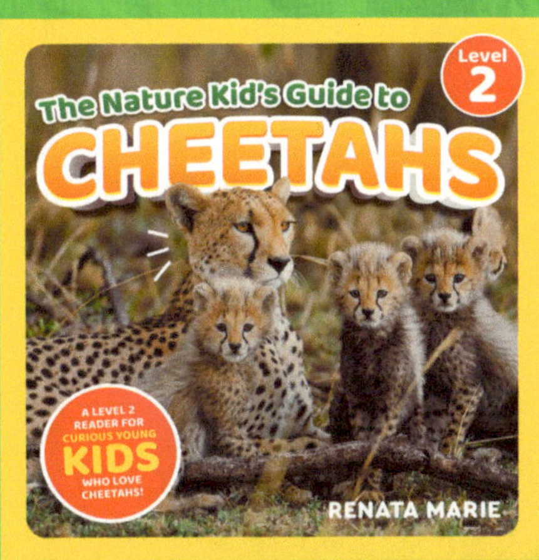

Cheetahs

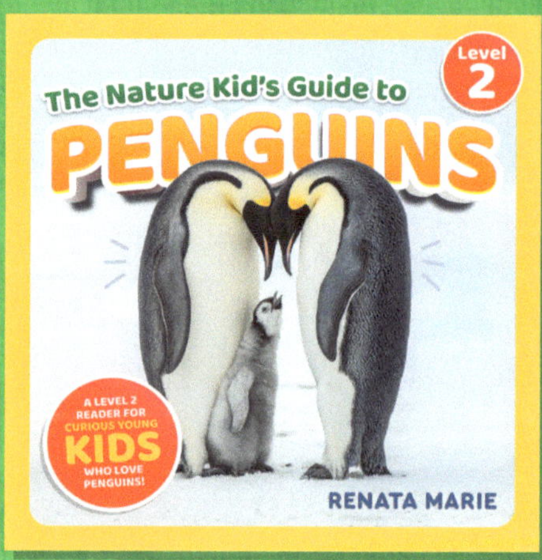

Penguins

Lions

Owls

Meerkats

Visit NatureKidsPublishing.com to Learn More!

www.ingramcontent.com/pod-product-compliance
Lightning Source LLC
Chambersburg PA
CBHW042123030726
47599CB00002B/329